Contents

AF335193

Each page has a title telling you what it is about.

Can I afford it?

MF2.5

January						
Su	M	Tu	W	Th	F	Sa
		1	2	3	4	5
6	7	8	9	10	11	12
13	14	15	16	17	18	19
20	21	22	23	24	25	26
27	28	29	30	31		

February						
Su	M	Tu	W	Th	F	Sa
					1	2
3	4	5	6	7	8	9
10	11	12	13	14	15	16
17	18	19	20	21	22	23
24	25	26	27	28		

Kim saves £27 each Friday, starting on 4th January.
She wants to buy one of these things.
Say if she can afford the item. If she can,
how much will she have left?

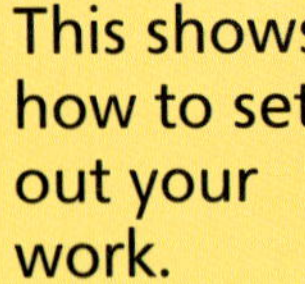

This shows how to set out your work.

Instructions look like this. Always read these carefully before starting.

1 This camera on 18th January?

2 This mobile on 1st February?

3 This TV on 10th February?

4 This Blu-ray player on 21st February?

5 This laptop on 1st March?

6 This sound system on 15th March?

These are Rocket activities. Ask your teacher if you need to do these questions.

I can work out a simple budget

Read this to check you understand what you have been learning on the page.

Airport prices

This is what each company charges for airport parking.

BCP	Kwik Parking	Supa Parking
£7·50 per day	£50 per week + £9 per day for additional days	£195 for any length of parking up to 5 weeks

Work out which company offers the best deal for each person. Use a calculator if you want to.

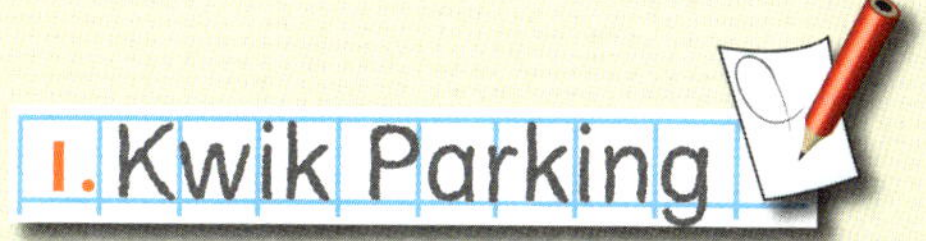

1 

2

3 

4

5 

6   Describe when it is best to use Supa Parking.

I can work out which is the best deal for different situations

Phone calls

Here are some phone companies' call charges.

Cheap Talk

Only 24p per minute at any time.

First minute free!

Fast-chat

21p per minute at any time.

Chat-chat

11p per minute cheap rate,
30p per minute between 9 am and 6 pm.

Work out the cost each company would charge for:

1   A 5 minute call at 9 pm.

2   A 7 minute call at 7 am.

3   An 8 minute call at 11 am.

4   A 12 minute call at 2 pm.

5   A 15 minute call at 5 pm.

6   Fast-chat charges £10·60 per month for line rental on top of the call charges. VAT at 20% is also added. Mrs Brown had 60 minutes of call charges this month. What was her final bill with Fast-chat?

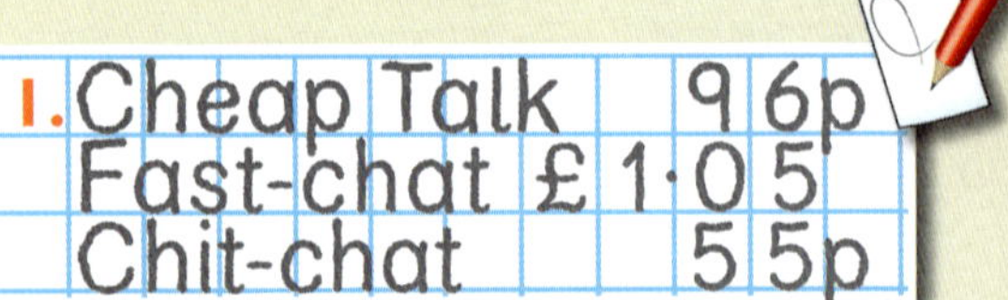

I can work out deals and compare them

Prices

Work out the price per 100 g.

1. 7 5 p per 1 0 0 g

1 £3·00
400 g

2 £2·00
200 g

3 £7·20
1 kg

4 90p
500 g

5 £5·00
5 kg

6 £1·50
1·5 kg

Work out the price per 100 ml.

7. 3 0 p per 1 0 0 ml

7 90p
300 ml

8 £1·80
2 l

9 60p
500 ml

10 £1·10 
1 l

11 £3·00
3 l

12 £5·50
5·5 l

Talk to your partner about which cola is best value for money.

I can work out the price per 100 g or 100 ml

Value for money

Work out which in each set is the best value for money.

1

a — 6 pack £2·82

b — 9 pack £4·14

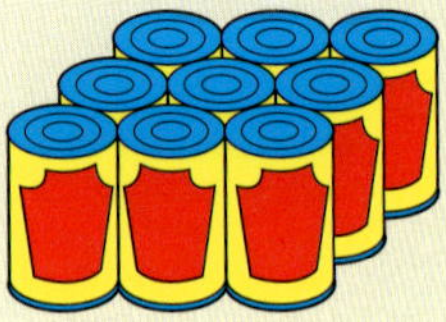

c — 2 pack £1·10

2

a — 1 pack 85p

b — 4 pack £3·04

c — 10 pack £8·08

3

a — 500 g £3·00

b — 100 g 78p

c — 350 g £2·24

4

a — 2 l £1·39

b — 1 l 89p

c — 5 l £3·75

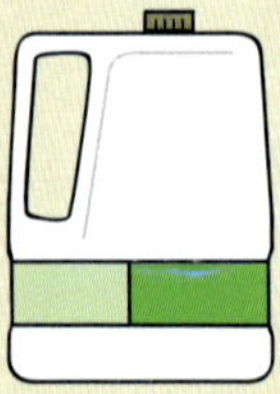

5 When buying some milk, why might Mrs Ahmed choose not to buy the pack that is best value for money?

Make up your own value for money puzzle for your partner to solve.

I can work out value for money

Which is cheaper?

Write which is cheaper and say by how much.

1 a £14·79   b £13·99

2 a £6·75   b £7·59

3 a £11·09   b £9·87

4 a £36   b £34·63

5 a £2·83   b £3·17

6 a £25·30   b £18·77

These items are discounted. How much cheaper is each one now?

£9·92

NOW HALF PRICE!!!

£3·48

SALE – ONE QUARTER OFF!!!

If you had £35 which of the items on the page would you buy?

I can say which is cheaper and by how much

Can I afford it?

Write yes or no to say if there is enough money to pay for each item.

If yes, say how much money you will be left with.
If no, say how much more you would need.

1

£7·63

2

£32

3

£89·50

4

£51·74

5 Amy has £180 in the bank. Can she afford to buy all four items?
Say how much is left, or how much is needed.

I can compare how much I have with the cost of items

Can I afford it?

January						
Su	M	Tu	W	Th	F	Sa
		1	2	3	4	5
6	7	8	9	10	11	12
13	14	15	16	17	18	19
20	21	22	23	24	25	26
27	28	29	30	31		

February						
Su	M	Tu	W	Th	F	Sa
					1	2
3	4	5	6	7	8	9
10	11	12	13	14	15	16
17	18	19	20	21	22	23
24	25	26	27	28		

Kim saves £27 each Friday, starting on 4th January.
She wants to buy one of these things.
Say if she can afford the item. If she can,
how much will she have left?

1 This camera on 18th January?

2 This mobile on 1st February?

3 This TV on 10th February?

4 This Blu-ray player on 21st February?

5 This laptop on 1st March?

6 This sound system on 15th March?

Kim buys the camera on 18th January and continues
to save £27 per week. Will she be able to afford a
£244 games console on 27th March?

I can work out a simple budget

Credit card statements

Sam's credit card statement shows what he has bought with his credit card and the payments he has made.

Calculate:

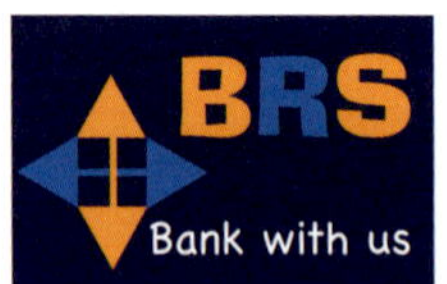

1   Sam's total debits for this period

2   the total before interest

3   the amount of interest

4   the total amount he owes.

5   If Sam had paid for these things on a debit card he wouldn't have paid interest. How much would he have saved?

Credit card statement

BRS Bank with us

Mr SF McLaughlin   Account number 3856783

Due Date 31st July   Minimum Due £31

Reference number	Date	Transaction	Debits	Payments
005404	27/04	CA Airlines	£102·14	
356483	28/04	AB Stores	£45·21	
000132	30/04	Payment – thank you		£31·00
094743	07/05	RB Motors Ltd	£33·72	
257492	09/05	Supastore	£144·93	

Total debits	
Payment	- £31·00
Total before interest	
+ 15% interest	
Total amount owed	

Why do you think Sam used a credit card instead of a debit card? Discuss this with your partner then write your answer.

I can understand credit card statements

Money words

Here are some words related to money.

loan **account** **save** **borrow**

statement **interest** **pension**

Copy these sentences and write which word or words are missing from each one.

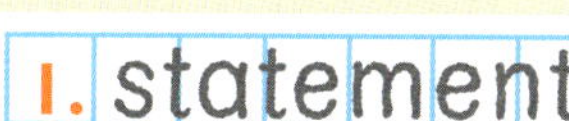

1 Each month Kim receives a ____________ to show how much she has in the bank.

2 Carl wants to buy a car but doesn't have enough money. He decides to _________ some so that he can buy it now.

3 Each month Dev pays money into a ____________ so that he will get regular money when he is retired.

4 When paying back a ____________ Susie has to pay extra in _________ charges.

5 Paige decides to ____________ some money and opens a new _________ at the bank to put it in.

Write your own sentences for each of these words.

budget **mortgage** **overdraft**

Savings

Part of this statement for a savings account has been stained.

Savings account statement

Karen Lines   Account number 47636483

Date 31st September

Date	Transaction	Withdrawals	Deposits	Balance
01 Sep	Existing balance			£137·26
17 Sep	Cash machine withdrawal	£50·00		£87·26
21 Sep	Cheque deposit		£25·00	£112·26
27 Sep	Cash machine withdrawal	£40·00		£72·26
28 Sep	Withdrawal at bank	£55·00		
30 Sep	Interest payment		£6·63	

How much was in Ms Lines' account on:

1.   5th September?   2.   24th September?

What transaction happened on:

3.   21st September?   4.   27th September?

Calculate the balance at the end of the day on:

5.   28th September.   6.   30th September.

7.   Clive has £120 in a savings account. He doesn't withdraw or deposit any money but gets 5% interest each year. How much interest will he get in the first year?

If Clive keeps his account going for 3 years, and never withdraws or deposits anything, how much will he have in it by the end of that time?

I can interpret statements and talk about interest payments

Balances

Urvi had a balance of £82·25 in her bank account on 1st July.

1 Use these receipts and paying-in slips to work out her balance on 20th July.

Sunny's Supermarket
03/07
TOTAL SALE:
£32·10
Payment by debit card
**** **** **** 1235
THANK YOU

Fred's Super Saver Fuel
07/07
TOTAL SALE:
£39·05
Payment by debit card
**** **** **** 1235
THANK YOU

Paying-in slip
Date: 10th July

Cashier's Stamp

NCBA SECURITY

NCBF SECURITY 20978 1 1xPk500, 01/11

Amount paid in
£250·00

Number of cheques: 1

NATIONAL BANK
Date: 15/07
YOU HAVE WITHDRAWN:
£150·00
AT HIGH STREET
TRANSACTION REF: 00285
Account: 1235

The Shop on the Corner
16/07
TOTAL SALE:
£41·90
Payment by debit card
**** **** **** 1235
THANK YOU

The Post Office
19/07
TOTAL SALE:
£22·50
Payment by debit card
**** **** **** 1235
THANK YOU

2 What were Urvi's outgoings during this time?

3 How much did she pay into her account in total during this time?

On the last day of July, Urvi had a balance of £84·50 in her account. Write some possible transactions she might have had between 20th July and 31st July.

I can work out balances and write a statement

Some people bought items on a plane.
They could pay in pounds or in Euros.
The exchange rate was £1 = 1·16 Euros.

£9·50

£4·75

£23·80

£13·25

£15·80

£8·40

£54·65

£48·28

Which item did each person buy?

1.James bought the
£9·50 box of chocolates

1   James paid €11·02.

2   Chloe paid €15·37.

3   Susie paid €5·51.

4   Raz paid €27·61.

5   Deepa paid €56·00.

6   Kit paid €18·33.

7   Fiona paid €63·39.

Jamie had €239 left at the end of his holiday and changed it back to pounds, using the same exchange rate. How much did he get back rounded to the nearest pound? Use a calculator if you want.

I can use exchange rates to do conversions

Currencies

Copy the table and extend it up to £10. Draw a line graph marking the horizontal axis up to £10. Use the graph to convert the prices!

x-axis (£)	1	2	3	4
y-axis (Rupees)	80	160	240	

1   800 R

2 240 R

3   440 R

4   560 R

5   480 R

6   360 R

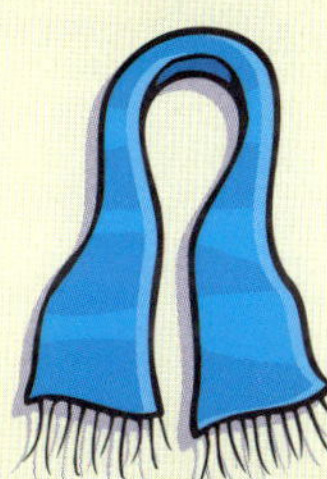

7   760 R

8   800 R

9   200 R

How much will Hilda get in rupees for:

10 £3   **11** £5   **12** £10   **13** £2·50

14 £15   **15** £7·50   **16** £40

Explore the value in pounds if you have 1 million rupees. Estimate first with your partner.

I can use a table and graph to help me convert between currencies

Currencies

On one day, the currency exchange rates were:

1 GBP = 1·14 Euros	1 Euro = £0·88
1 GBP = 1·63 US Dollars	1 US Dollar = £0·61
1 GBP = 132 Japanese Yen	100 Japanese Yen = £0·75

Some people changed currencies at an exchange bureau.
Work out how much each person got using the rates above.

1   Mr D changed £200 to US Dollars.

2   Mrs R changed £150 to Euros.

3   Ms J changed £500 to Japanese Yen.

4   Miss S changed 250 US Dollars to pounds.

5   Mr B changed 300 Euros to pounds.

6   Mr B changed 120 000 Japanese Yen to pounds.

7   Ms H changed £350 to Japanese Yen.

8   Mr Y changed £190 to Euros.

9   Lord F changed 2500 Euros to pounds.

Add a new exchange rate to the table and write some questions for your partner to answer.

I can use exchange rates to do currency conversions

Profit or loss?

Say whether each enterprise made a profit or loss and say by how much.

1. Jack's class made 150 badges and sold them for 80p each. It had cost £45 for the safety pins, £20 for the cardboard and £10 for the paints.

2. Kia's class made 225 cakes and sold them each for 25p. It had cost £15 for the flour and eggs, £13 for sugar, £14 for butter, £12 for coloured icing and £6 for cherries and cake cases.

3. Kyle's class bought 200 white plates for £1 each. Using paint costing £24, they decorated them and then sold each plate for £3·20. They only managed to sell 80 of the plates.

4. Fi's class ran a talent show. 67 people entered the competition, each paying a 40p entry fee. They awarded a £25 prize to the winning act.

5. Carl's class paid £55 for a local guitarist to give a concert. 44 people came to the concert, each paying a £1·80 entry fee which included one free drink. The drinks cost £24·50.

Invent an enterprise project like these where the class made a profit of £15.

Enterprise decisions

Lucy's class is making chocolate cookies to sell. This list shows how much they paid for ingredients.

Expenses:
5 bags of flour at £2·50 per bag.
2 bags of sugar at £1·85 per bag.
4 packs of butter at £1·25 per pack.
6 bars of chocolate at 90p per bar.
4 bags of chocolate drops at 85p per bag.

1 What is the total of all their expenses?

They made 240 cookies in total. They think about charging 20p per cookie.

2 If they charge 20p for one cookie, how many do they need to sell to make a profit?

3 If they sell all the cookies at 20p each how much profit will they make?

They finally decide to sell the cookies for 25p each.

4 How many do they need to sell at 25p to make a 50% profit?

5 If they sell all the cookies, what percentage profit will they make?

6 In the end they sold 214 cookies and gave the profit to charity. How much did they give to charity?

 I can do enterprise calculations

Enterprise report

Now that you have completed an enterprise project, you have been asked to produce a report on how it went, for a TV show.

Write a script for the report and video it.

In your report include:

- what the enterprise was

- who was involved

- what roles each of you took

- how you set it up

- what happened

- what went well

- whether there were any difficulties

- whether you made a profit or loss

- how happy you were with the outcomes

- what happened to any profit.

Think of different ways to make the report interesting for the audience.

You could use charts or pictures, interviews with people, show the enterprise at work and so on.

Think of some advice for other children who are going to do a similar thing.

I can report on how successful my enterprise project was

Grouping data

Look at the table. It gives information about how many hot drinks people drank on one day.

Cups	1	2	3	4	5	6	7	8	9	10	11	12
Tallies	I	II	IIII	IIII	IIII	III	III	I	II	I	I	I

1 Group the data in the following way: 1–2 cups, 3–4 cups, 5–6 cups …

2 Draw a frequency table.

3 Draw and label a bar graph.

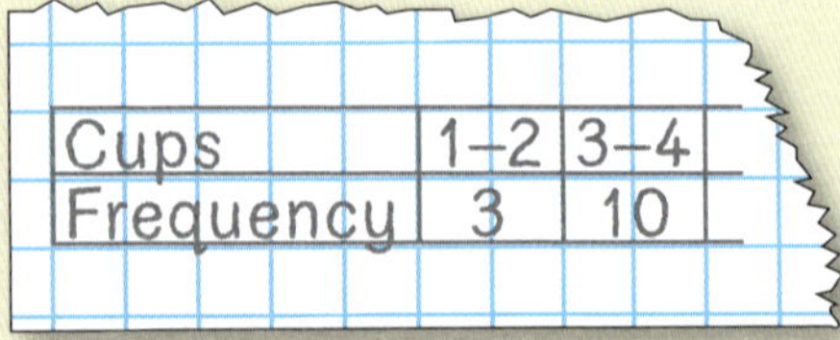

4 How many cups were drunk most frequently?

5 How many cups were drunk least frequently?

6 How many people drank more than 6 cups?

7 How many people drank fewer than 3 cups?

8 How many people drank more than 2 and fewer than 9 cups?

9 True or false: half the people asked drank up to 6 cups per day.

If you added the adults in your house to this information, would it affect the most common frequency?

I can group data

Grouping data

Look at the graph. Use it to answer the questions.

1 How is the information grouped?

2 Which number of packets is the most common?

3 How many children ate 21–25 packets of crisps in May?

4 How many children ate 1–5 packets?

5 How many children took part in the survey?

6 For packets of crisps, what is the range?

How many children ate:

7 more than 20 packets?

8 fewer than 11 packets?

9 between 11 and 25 packets?

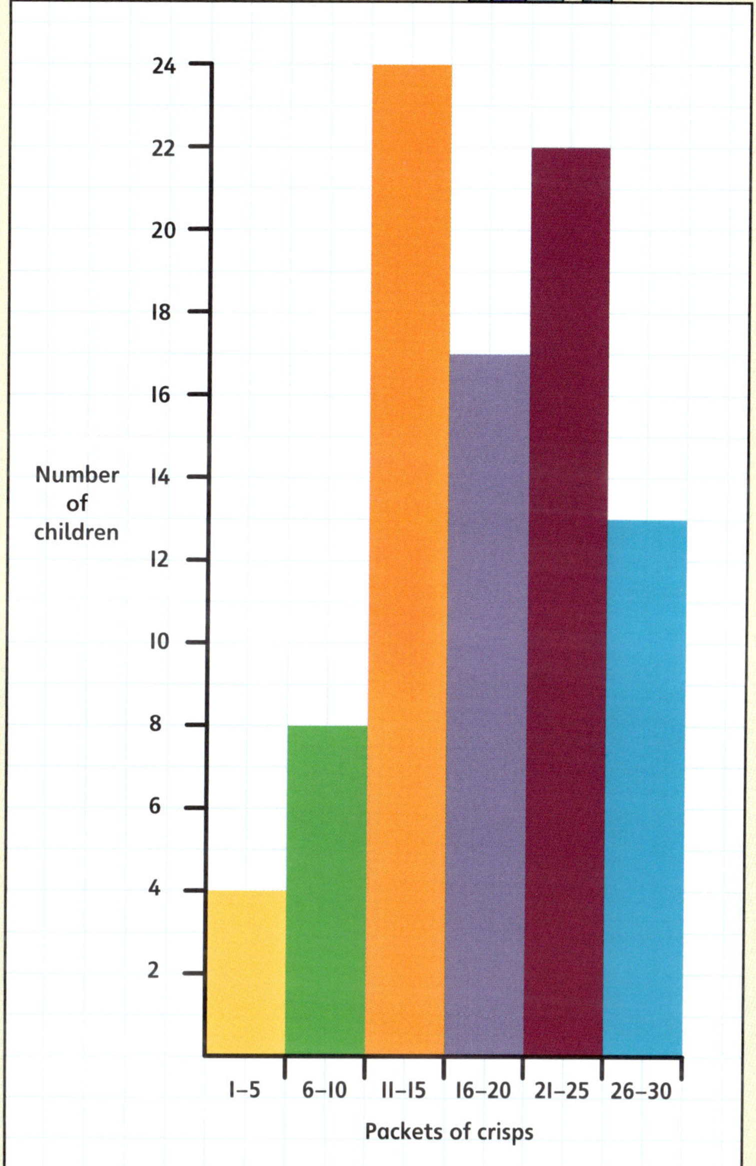

Crisps eaten by Class Z in May

Discuss with your partner the number of packets of crisps you eat each month. Which bar(s) on the graph would change if you were to add your data?

I can interpret grouped data

Bar graphs

Weight of rubbish found

Bin A in shopping mall

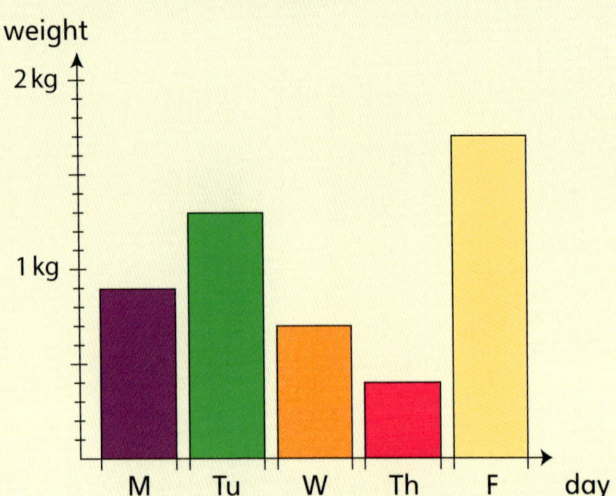

Bin B in bus station

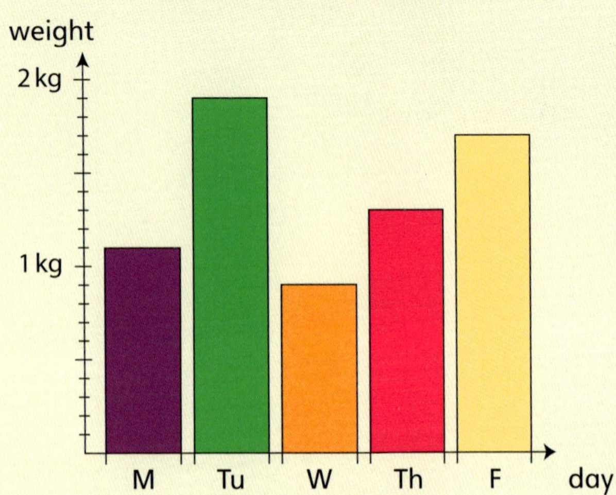

How much rubbish was found:

In Bin A on: **1** Wednesday? **2** Friday? **3** Monday?

In Bin B on: **4** Tuesday? **5** Thursday? **6** Friday?

7 On which days was more rubbish found in the bus station than in the mall?

How much less rubbish was found in Bin A than in Bin B on:

8 Monday? **9** Thursday?

What was the total rubbish found in both bins on:

10 Thursday? **11** Tuesday?

What was the total rubbish for the 5 days in:

12 Bin A? **13** Bin B?

14 Estimate the amount of rubbish in each place on Saturday and Sunday. Explain your estimates.

What could be the reasons why there is a greater weight of rubbish on Tuesdays and Fridays?

I can interpret bar graphs

Bar line graphs

How many dogs had:

1 3 puppies?

2 5 puppies?

3 9 puppies?

Woofshire Dog Breeders 2012

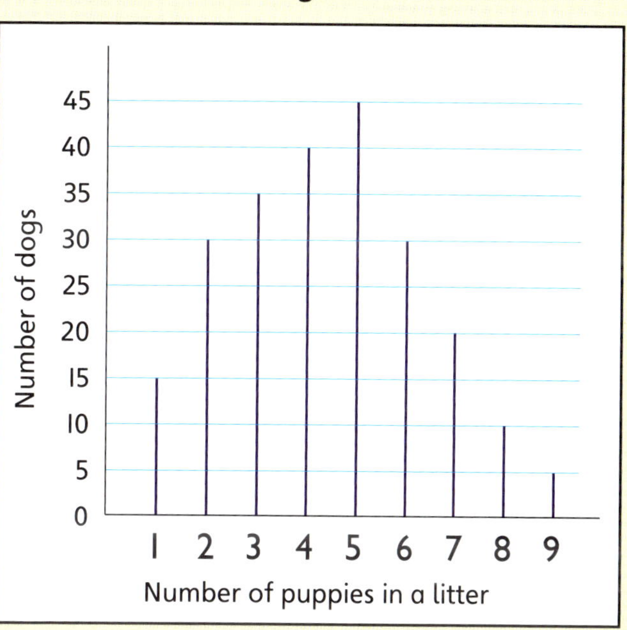

4 What was the smallest litter size?
How many dogs had litters of this size?

5 What was the largest litter size?
How many dogs had litters of this size?

How many dogs had:

6 more than 5 puppies?

7 fewer than 4 puppies?

8 between 2 and 7 puppies?

9 Make up some questions about this graph for your partner to answer.

Choose an animal. Draw a graph to show its litter size.

I can interpret bar line graphs

Bar line charts

The calendar shows the temperature recordings for 28 days in October.

Monday	Tuesday	Wednesday	Thursday	Friday	Saturday	Sunday
10°	12°	13°	15°	16°	15°	15°
13°	11°	14°	13°	12°	13°	13°
12°	14°	13°	14°	11°	12°	12°
13°	12°	11°	13°	14°	13°	13°

I   Use the data in the calendar to draw a frequency chart.

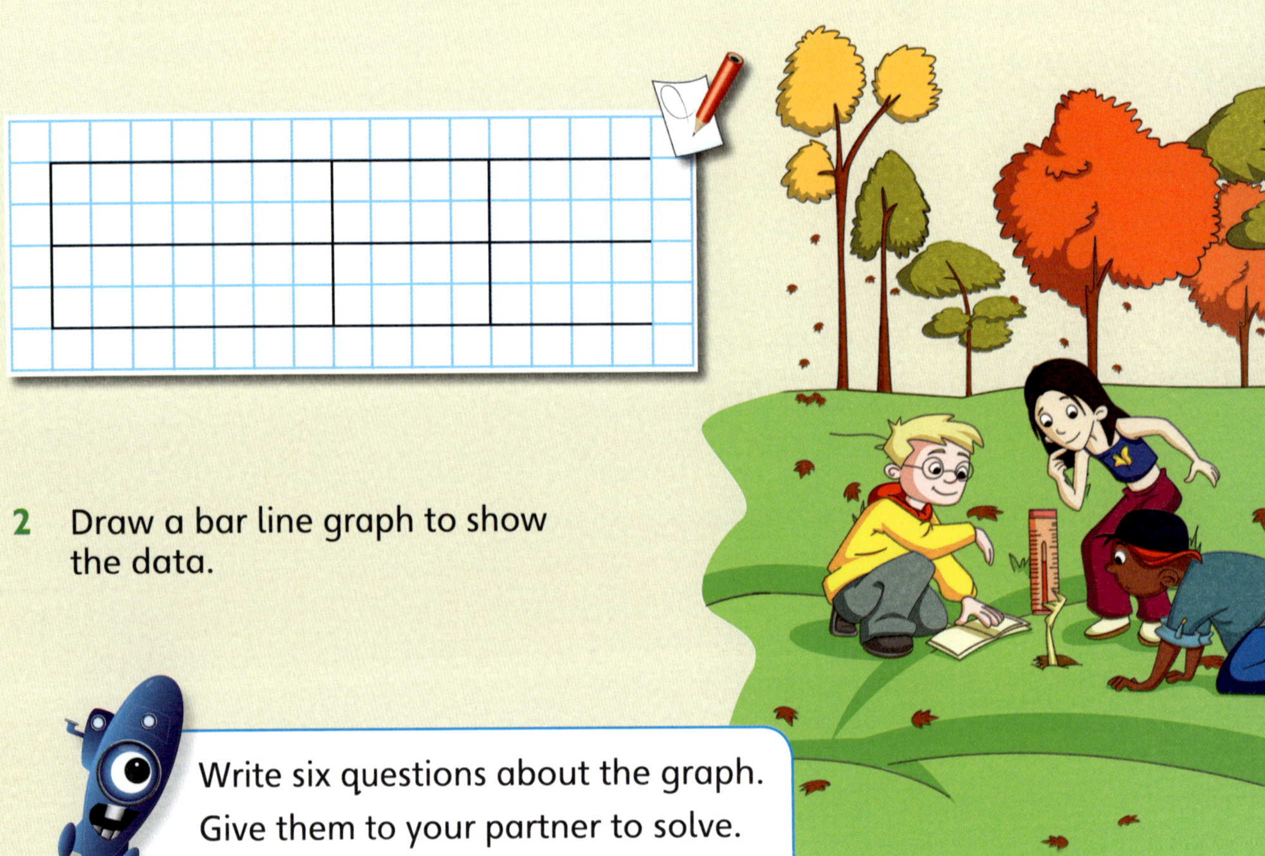

2   Draw a bar line graph to show
the data.

I can make decisions about suitable scales when drawing a bar line graph

Always, sometimes, never

Say whether each statement is:
always true; sometimes true; never true.

1   The day after Sunday is Tuesday.

2   The day after Wednesday is Thursday.

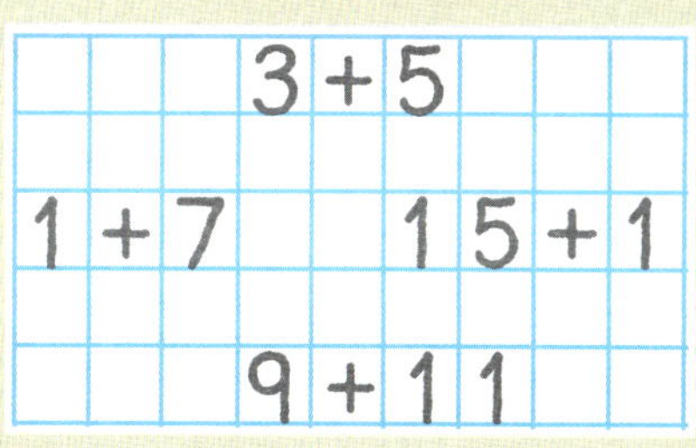

3   An hour has 60 minutes.

4   A month has 30 days.

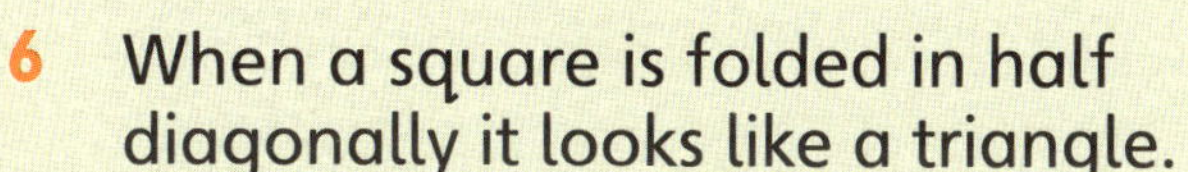

5   When a triangle is folded in half
    it looks like a square.

6   When a square is folded in half
    diagonally it looks like a triangle.

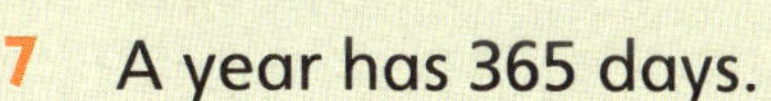

7   A year has 365 days.

8   The total of two odd numbers is even.

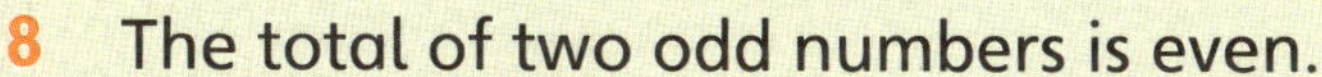

Make up some statements and ask your partner
to say whether they are: always true; sometimes
true; never true.

I can say whether something is always, sometimes or never true

Likelihood

Say whether each event is:
impossible; unlikely; likely to happen.

1 Someone in your family will win the lottery.

2 You will see a car today.

3 It will rain tomorrow.

4 You will go to the moon this week.

5 Someone in your class will be away tomorrow.

6 Your teacher will sing a song tomorrow.

7 You will meet a crocodile on your way to school tomorrow.

8 It will snow this week.

Write an event that is:

9 likely

10 unlikely

11 impossible

I can say how likely something is to happen

Likelihood

Draw a table like this. Decide how likely each event is and write
the number in the correct column of the table.

likely	unlikely	impossible
		1

1. I will land on Mars tonight.
2. I will go to bed late tonight.
3. I will walk more than 10 steps today.
4. I will find a four-leaf clover today.
5. I will watch television tonight.

Draw a table like this. Decide how often each event happens and write
the number in the correct column of the table.

always	sometimes	never

6. I go to sleep at night.
7. The moon shines in my window.
8. I have sweet dreams.
9. I wake up and get up.
10. I take the tiger for a walk.
11. I eat ice-cream for breakfast.

Make up more statements for each
column of your two tables.

I can say how likely something is to happen

Questionnaires

Work with your partner. Choose one of the questions below to find out about.

Devise a questionnaire or recording sheet.

Is it true that the children in our school would prefer to start school one hour earlier and go home one hour earlier?

Are girls more likely than boys to play a musical instrument?

Which is the most common day of the week for children in our class to have their birthday this year?

Are people who own pets more likely to be vegetarian?

Think about...

Are the questions clear?

What might the answers be?

Do you need to provide answers for people to choose from?

Have you got enough space for recording the information?

Is the sheet quick to fill in?

I can write my own questionnaires

Questionnaires

Here are two questionnaires. One is a much better questionnaire than the other.

I Write at least five reasons to say why the second is better.

> Do you play games on a computer?
> Do you like it?
> How often do you do it?
> For how long do you play?
> What is your favourite game?

I Do you play computer games? ☐ Yes ☐ No
(If no, stop the survey now)

2 How often do you play them?

Every day ☐
Most days ☐
Once or twice a week ☐
Less than once a week ☐
Other ....

3 About how many hours a week do you play computer games?

More than 28 hours (more than 4 hours a day) ☐
Between 14 and 28 hours (about 2–4 hours a day) ☐
Between 7 and 14 hours (about 1–2 hours a day) ☐
Less than 7 hours (less than 1 hour a day) ☐
Other ....

4 What are your favourite types of game?

adventure ☐ racing ☐ sports ☐ music ☐
action ☐ exercise ☐ puzzle ☐ arcade ☐

Other ....

5 What is your favourite game? ....

I can understand what makes a good questionnaire

Databases

These flowers have different central colours, with different numbers of petals, with different shaped petals and different coloured petals.

a   b   c   d

e   f   g   h

i   j   k   l

I Make a database to record this information about the flowers.
Use these four field headings:

Central colour, Petal colour, Petal shape, Number of petals

2 Write some statements about the information in your database.
For example, '4 flowers have yellow petals'.

I can make my own database and sort the information

Databases

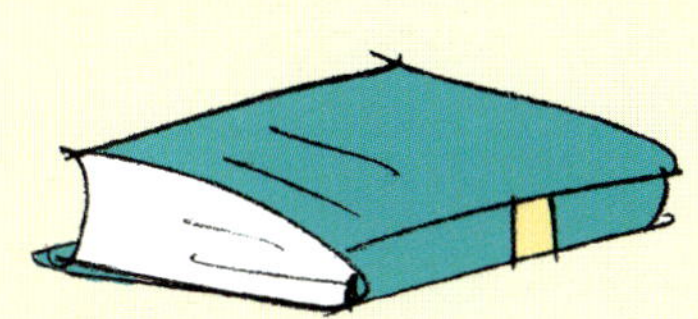

1 Choose about 25 books from your library. You are going to make a database about them.

Choose five different fields to describe them. You need at least one field with a numerical answer and at least one where the answer is 'yes' or 'no'.

You could choose from:

- title

- author

- the year the book was published

- number of pages

- type of cover

- with pictures

- fiction

- suitable age range

- in colour

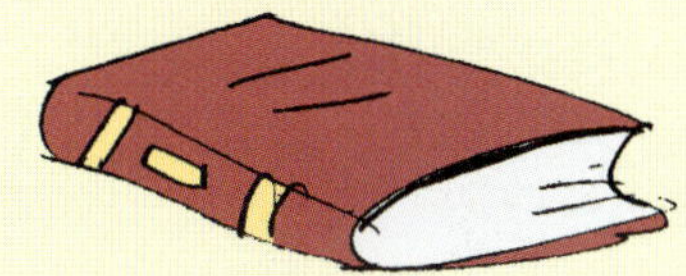

2 Make a paper or computer database to record data in your five fields.

3 If you made a computer database, sort your database in different ways.

Think about sorting the data according to each field. Which produces the most interesting information? Why?

4 Write at least four statements about the information in your database.

I can make my own database and sort the information it contains

Sorting

Below is part of a database showing shapes in a set. All the shapes are prisms.

- The end-faces are a circle, a square or a triangle.
- Each shape is red or yellow.
- Each shape is large or small.
- Each shape is thick or thin.

Shape of end-face	Colour	Size	Thickness
triangle	yellow	large	thick
square	red	small	thick
circle	yellow	large	thin
triangle	yellow	small	thin
circle	red	large	thick
square	red	small	thin
circle	yellow	small	thick
triangle	red	large	thick

1 Make a database of this data on a computer.

To help you complete the next tasks, sort your data according to different fields.

2 There are 24 different shapes in the complete set. Add entries to your database to show the remaining shapes.

3 Write four statements about the information in your database.

What if there were blue shapes as well? How many new entries would you need to make?

I can make my own database and add new entries

Pie charts

Car colour

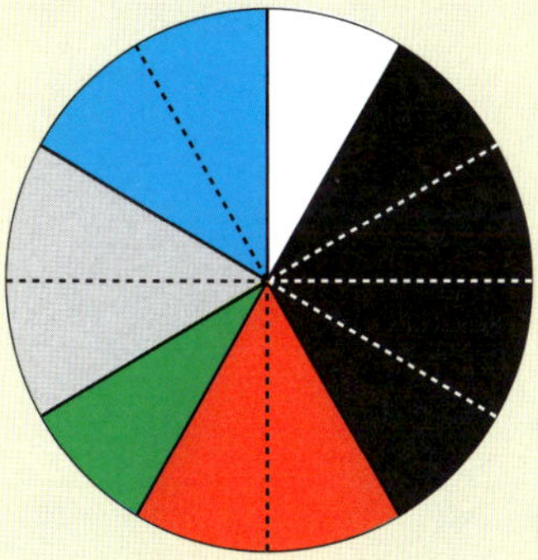

120 cars were surveyed

What fraction of cars surveyed were:

1. white?
2. blue?
3. green?
4. black?
5. red or green?
6. white or black?
7. not blue or green?
8. one of the colours of the rainbow?

1. $\dfrac{1}{12}$

How many cars were:

9. red?
10. black?
11. grey or white?
12. green or blue?
13. red or white?
14. yellow?

9. 20 cars

This chart shows favourite drinks. Discuss what each slice might represent, e.g. pink = juice. If 16 friends are surveyed, how many would be in each category?

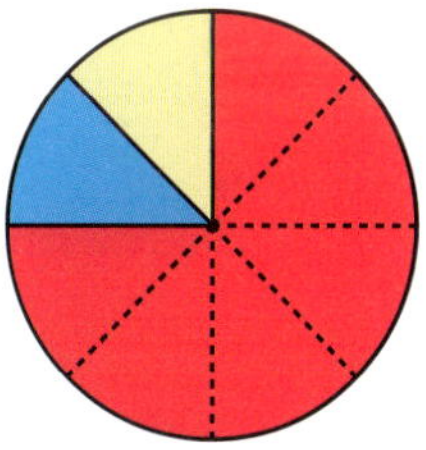

Pie charts

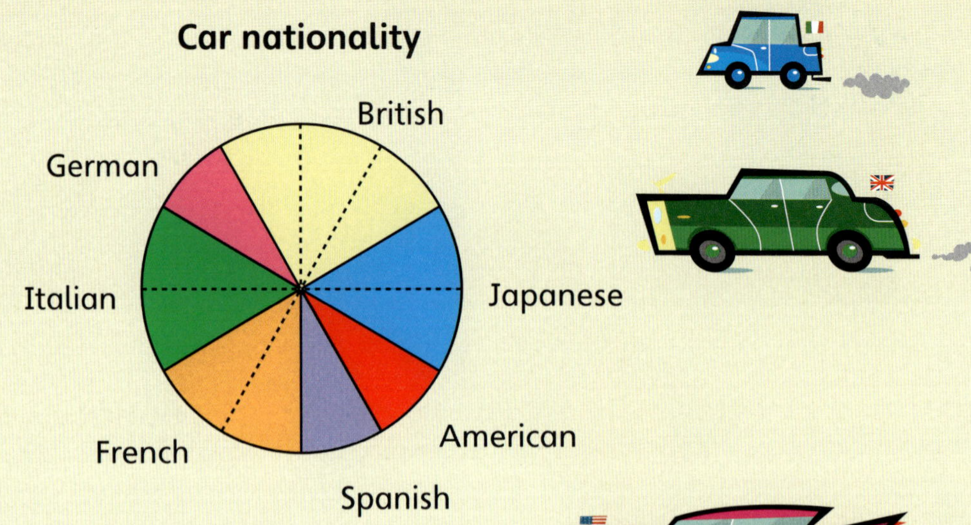

What fraction of cars surveyed were:

1 British?
2 French?
3 not British?

4 Japanese?
5 French or Spanish?
6 British or German?

How many cars were:

7 French?
8 American?
9 non-European?

10 not British?
11 European?
12 Japanese?

13 Write the names of 12 children in your class. Look at the letters in their names then complete this table:

Number of letters	2–3	4–5	6–7	8–9	> 9
Number of names					

Use the table to draw a pie chart.

I can interpret information in a pie chart

Pie charts

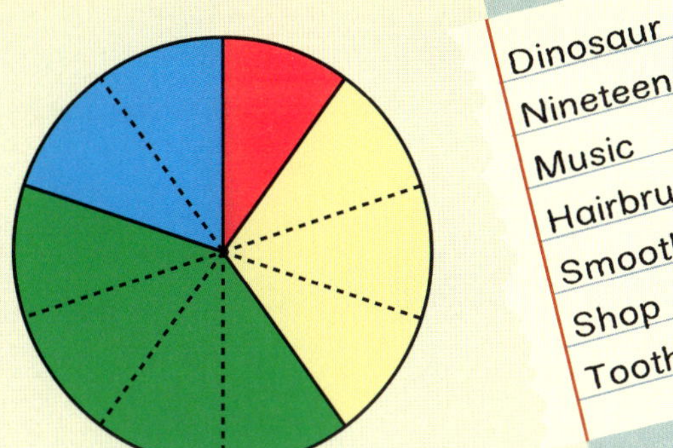

Lorna picked 20 words from her dictionary at random and counted their letters. She made a table then a pie chart to show her data.

1 Copy and complete Lorna's table.

Word length	1–3	4–6	7–9	10–12
Number of words				

Which slice could represent words with these numbers of letters?

2 1–3   **3** 4–6   **4** 7–9   **5** 10–12

How many words have:

6 6 or fewer letters?   **7** 10 or more letters?

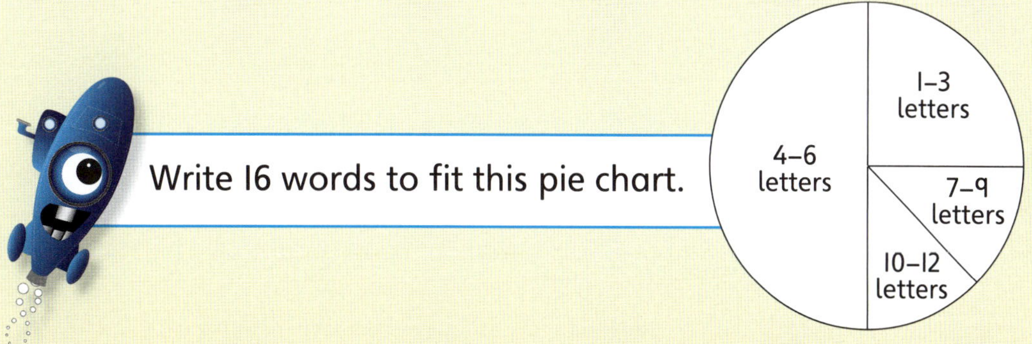

I can make links between data shown in a table and a pie chart

Line graphs

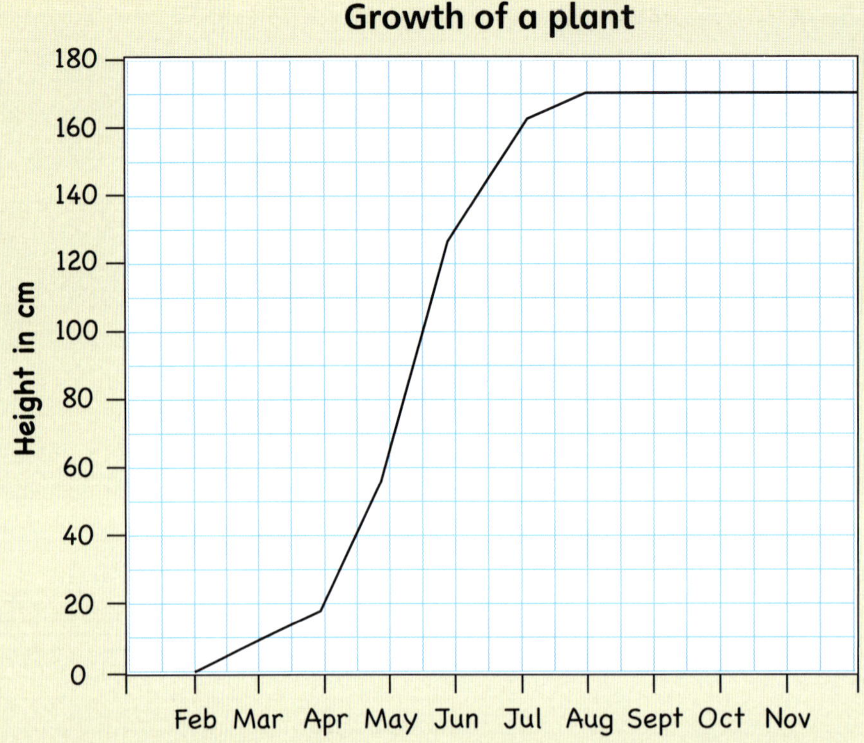

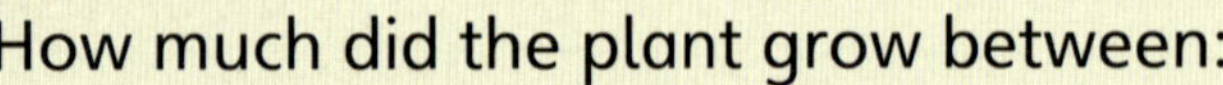

How much did the plant grow between:

1 Ist April and Ist May?   **2** Ist June and Ist July?

3 Ist September and Ist October?   **4** Ist February and Ist August?

5 In which month did the plant grow the most?

6 In which months did the plant not grow any taller?

7 Use the graph to help you write three facts of your own about the plant's growth.

Sketch a graph showing your probable height over the next few years. Write some questions about it for your partner to answer.

I can interpret line graphs

Line graphs

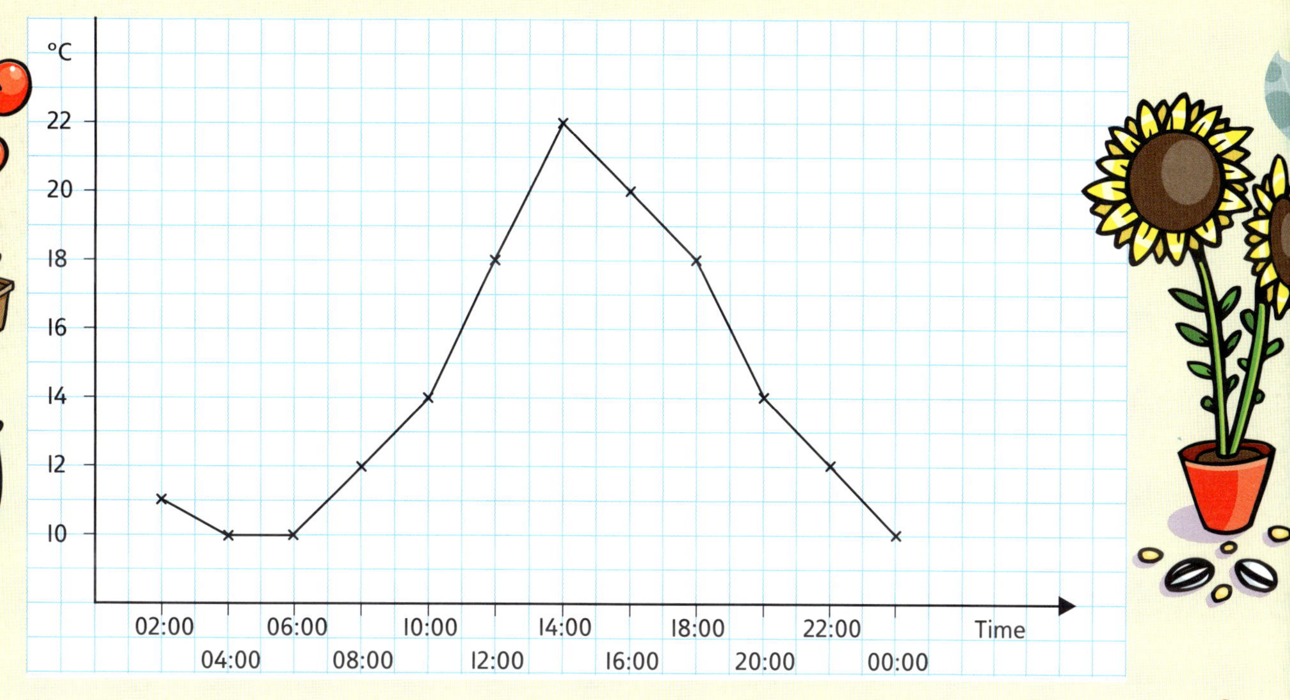

The temperature in the greenhouse was measured and recorded every 2 hours, starting at 02:00.

At what times was the temperature:

1 10°?   2 12°?   3 18°?   4 14°?   5 22°?   6 20°?

When is the temperature:

7 highest?   8 lowest?   9 half-way between 12° and 20°?

What is the temperature at:

10 04:00?   11 02:00?   12 20:00?   13 00:00?

What is the approximate temperature at:

14 07:00?   15 19:00?   16 13:00?   17 11:00?

Draw a graph giving the approximate temperatures in your classroom over the school day.

I can interpret line graphs

Line graphs

Level of water in the River Lagan in March

Date	1	3	5	7	9	11	13	15	17	19	21	23	25	27	29	31
Water level (cm)	85	85·5	86	88	87	89	93	94	94	93	90	89	87·5	89	90	89·5

The water level in the River Lagan was measured
and recorded every second day in March.

1 Draw a line graph to represent
this information.

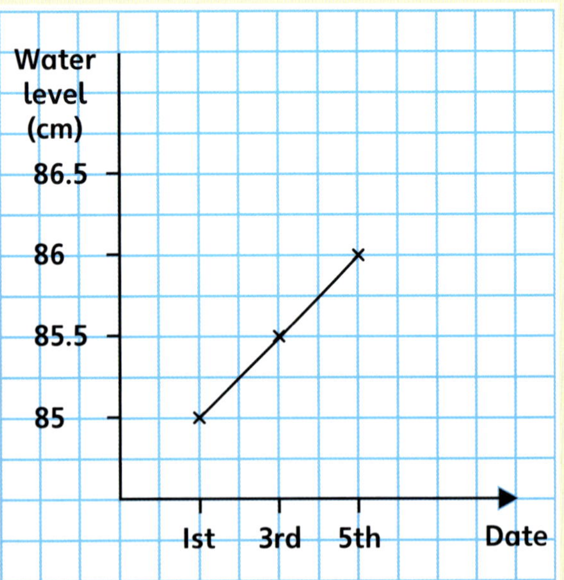

On what dates was the water level:

2 highest? **3** lowest?

4 half-way between highest and lowest?

What was the approximate water level on:

5 12th March? **6** 22nd March? **7** 28th March?

I can draw a line graph

Line graphs

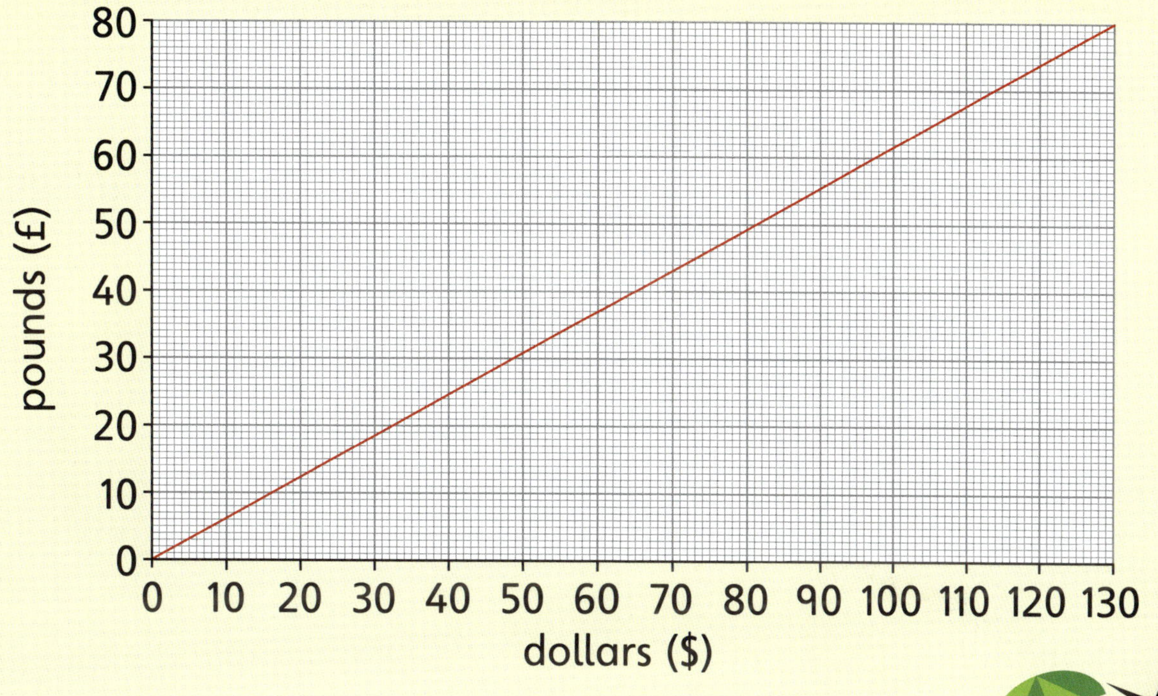

Read the graph to find out how many dollars you will get for:

1 £20 **2** £50 **3** £80 **4** £70 **5** £45

Read the graph to find out how many pounds you will get for:

6 $32 **7** $80 **8** $96 **9** $64 **10** $26

Find out the exchange rate for dollars and pounds today. Is it the same as the rate shown in the graph?

I can work out the value of any given point on a line graph

Probability

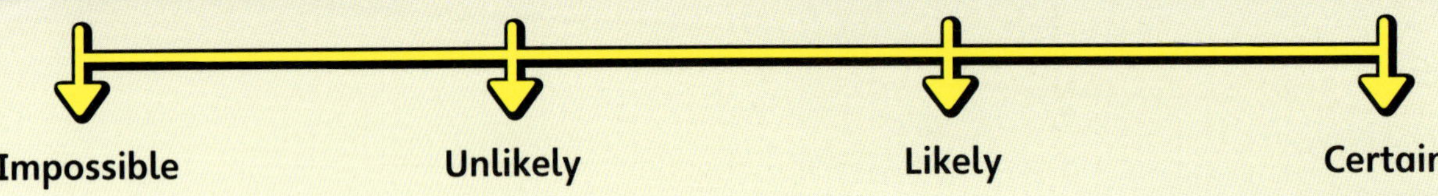

Choose one of the words above to suit each statement.

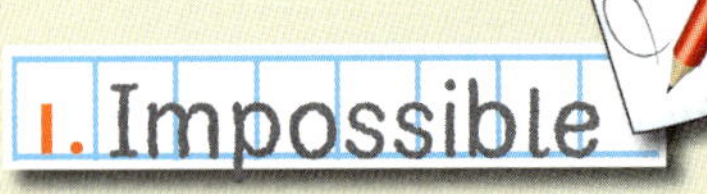

1 I will land on Mars tonight.
2 I will go to bed late tonight.
3 I will walk more than 10 steps today.
4 My toenails will grow 10 cm this month.
5 I will be younger tomorrow than I am today.
6 My skin will become purple overnight.
7 I will find a four-leaf clover today.
8 I will watch television tonight.

A coin is flipped 10 times. Write Impossible, Unlikely, Likely or Certain for each statement.

There were 10 heads.
We had 6 tails and 6 heads.
More than 4 times the coin was tails.
Heads and tails came up equal.
One-quarter of the throws came up tails.
6 throws were tails.
There were fewer than 3 heads.
8 were heads and 4 were tails.

I can say how likely things are to happen

Probability

Write one of these categories for each statement.

1 I will travel by rocket tomorrow. 

2 I will travel on an aeroplane this year. 

3 I will walk 20 miles this year.

4 I will use a computer today. 

5 I will eat potatoes this month. 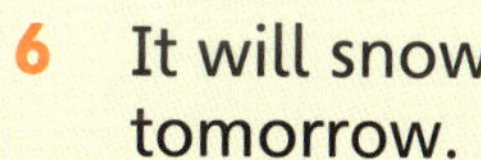

6 It will snow tomorrow.

7 I will go on a bus in June. 

8 I will see a dragon tomorrow.  

9 Write five statements of your own, one for each category.

Sarah takes one sock from her drawer.
What chance is there that it is:

10 yellow?

11 green?

12 blue?

13 not yellow?

14 pink?

15 either blue or green?

16 not green?

17 yellow, green or blue?

18 yellow or blue?

I can use a simple probability line

Probability

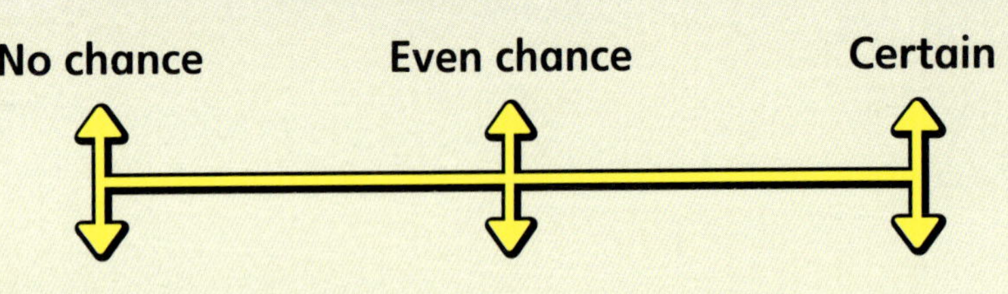

No chance   Even chance   Certain

Say if the chance of these events happening is:

- less than even
- more than even
- certain.
- no chance
- even chance

1. throwing a 6 on a dice

2. a coin landing heads

3. a dice landing on a number greater than 3

4. a card from a pack being red

5. a coin landing heads or tails

6. a card from a pack being a spade

7. a card from a pack being black

8. a dice landing on a number less than 6

9. a card from a pack being hearts or clubs

10. a dice landing on a 2 or 4

Use an 8-sided dice (numbered 1–8). Write an event that matches each of these: less than even, more than even, certain, no chance, even.

I can say whether things have more or less than an even chance

Medians

The children in H Class are collecting tokens.
Each time they bring some to class they record
how many they have added to their collection.

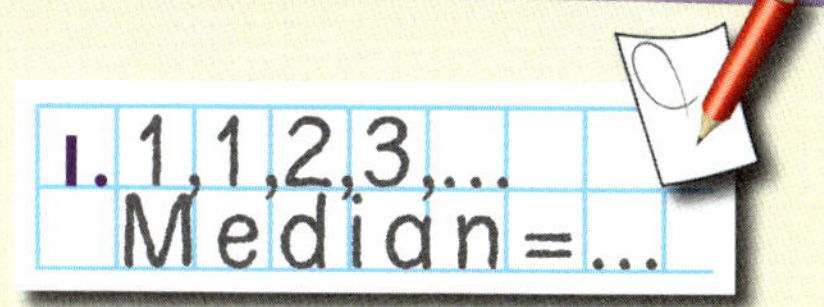

Calculate the median number collected by each child.

1 Tracey 4, 7, 3, 1, 2, 3, 4, 5, 1

2 Afram 1, 2, 1, 2, 3, 4, 5

3 Chrissie 2, 5, 3, 5, 2, 3, 4, 1

4 Hema 6, 7, 4, 6, 3, 8, 2, 2, 1, 9

5 Josh 3, 2, 1, 4, 2, 1, 3, 4

6 Nathan 5, 5, 5, 5, 3, 2, 5, 6, 7, 9, 5

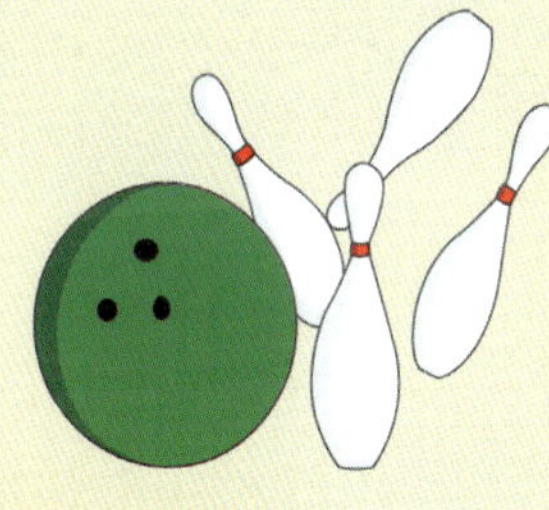

7 Jim goes 10-pin bowling. This shows how many
pins Jim knocked over in his first 11 bowls.

4	6	3	7	9	9	5	5	8	9	?

The median number knocked down in his first 11 bowls is 7.
What could his 11th score have been? Give three different answers.

Averages

Here are the goals scored by 7 teams over a number of games.
Calculate the mean for each team.

1 Reds: 5, 7, 4, 7, 7

2 Yellows: 8, 2, 7, 4, 4

3 Maroons: 3, 2, 5, 4, 4, 6

4 Greys: 8, 7, 7, 6

5 Whites: 6, 10, 8, 9, 9, 9, 5

6 Greens: 4, 3, 6, 5, 5, 7

7 Stripes: 6, 6, 8, 7, 9, 6

Six children took part in a football-bouncing competition. Here are the scores for how many bounces they managed each time.

Calculate the mean score for each child.

8 Dan   2, 1, 3, 7, 4, 6, 5, 8, 2, 1

8.

9 Ramesh 4, 3, 4, 2, 3, 4, 2, 3, 4, 2

10 Peter 6, 6, 3, 6, 3, 6, 7, 10, 7, 3

11 Cho   9, 9, 2, 8, 12, 10, 3, 3, 6, 10

12 Beth   4, 4, 8, 10, 4, 3, 6, 9, 2, 7

13 Guy   12, 15, 11, 12, 22, 13, 12, 9, 11, 15

14 Write the players in order, based on their mean scores.

Find some sets of scores where the mean is 6·5.

I can calculate the mean, including where the mean is a decimal

Averages

These are the temperatures in °C for the first 10 days in December. For each city, calculate (a) the mean, (b) the mode and (c) the median temperature.

1 London 7, 7, 6, 5, 5, 7, 8, 7, 7, 6

2 Belfast 8, 8, 6, 6, 4, 5, 6, 7, 6, 6

3 Athens 18, 18, 20, 20, 19, 18, 17, 19, 21, 21

4 Barcelona 12, 12, 13, 14, 13, 13, 12, 12, 12, 13

5 Dublin 7, 6, 6, 7, 6, 5, 6, 7, 7, 8

6 Bordeaux 10, 12, 12, 11, 10, 11, 12, 13, 12, 12

7 Singapore 28, 28, 27, 28, 28, 27, 26, 28, 29, 28

Sydney 38, 36, 35, 36, 35, 33, 34, 33, 33, 32

Calculate the missing number in each set of cards.

9
| 8 | 9 |
| [ ] | 6 |

mean = 7

10
| 5 | [ ] |
| 9 | 3 |

mean = 6

11
| [ ] |
| 11 | 6 |

mean = 8

12
| 17 | 18 | 13 |
| 16 | [ ] | 18 |

mean = 16

13 14
| 18 | | 20 |
| [ ] | 17 | 24 |

mean = 22

| 15 | 16 | 19 |
| 11 | [ ] | 18 | 17 |

mean = 18

Investigate the mean number of days per month in a year. Find the median and mode and compare the three averages. Do they change if it is a leap year?

Find out and show

Suppose someone wants to cook and sell pizzas for the school fair.

They would need to know which types of pizzas are most popular and which toppings people like best.

1 Plan how to collect this data for your school.
2 Design a questionnaire to use.
3 Collect the data. (You may just want to ask a sample of people.)
4 Organise the data and display it clearly.

Now if someone wants to make pizzas for the school, you can show them your data.

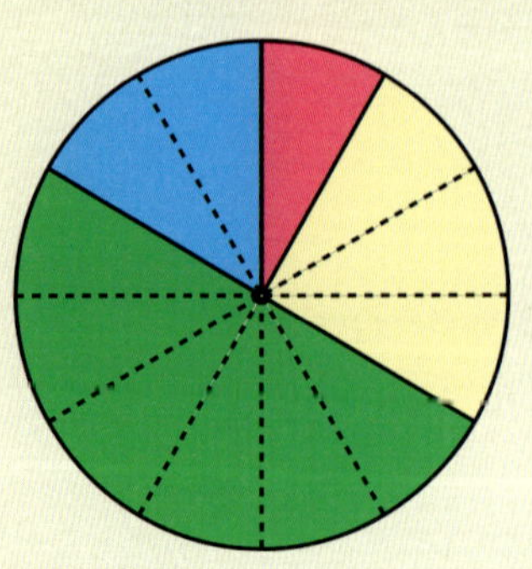

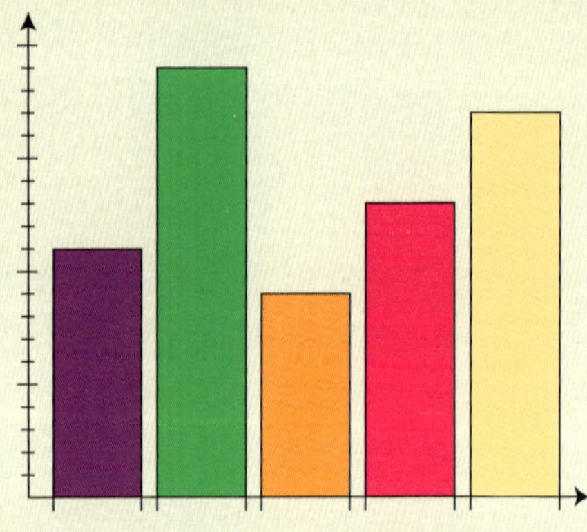

I can collect data and make decisions about how to display it

Find out and show

For the school fair, Claire wants to buy and sell cupcakes for a profit.

She wants to know how much they will cost her to buy and which kind are the most popular.

1. Find out what cupcakes are sold by an online supermarket, and how much they cost. If there are lots of cupcakes, you may have to pick just a few.

2. Collect this data and display it clearly.

3. Design a questionnaire to help you find out which of these cupcakes would be most popular.

4. Collect the data for children in your class. Organise it and display it clearly.

5. Suppose Claire asked your advice about what she should do. Now you have collected this data, which cupcakes would you suggest she buy?

6. How much should she charge for the cupcakes?

How many cupcakes should Claire buy if she was going to sell them at your school fair? What profit could she expect to make?

Author Team: Lynda Keith, Hilary Koll and Steve Mills
Consultant: Siobhán O'Doherty

Published by Pearson Education Limited, a company incorporated in England and Wales, having its registered office at Edinburgh Gate, Harlow, Essex, CM20 2JE. Registered company number: 872828

www.pearsonschools.co.uk

Text © Pearson Education Limited 2012

First published 2012

15 14 13 12
10 9 8 7 6 5 4 3 2 1

British Library Cataloguing in Publication Data
A catalogue record for this book is available from the British Library

ISBN 978 0 435 07742 6

Copyright notice
All rights reserved. No part of this publication may be reproduced in any form or by any means (including photocopying or storing it in any medium by electronic means and whether or not transiently or incidentally to some other use of this publication) without the written permission of the copyright owner, except in accordance with the provisions of the Copyright, Designs and Patents Act 1988 or under the terms of a licence issued by the Copyright Licensing Agency, Saffron House, 6–10 Kirby Street, London EC1N 8TS (www.cla.co.uk). Applications for the copyright owner's written permission should be addressed to the publisher.

Typeset by Debbie Oatley @ room9design and revised by Mike Brain Graphic Design Limited, Oxford
Illustrations © Harcourt Education Limited 2006–2007, Pearson Education Limited 2011
Illustrated by Matt Buckley, Seb Burnett, John Haslam, Anthony Rule, Chris Winn, Piers Baker, Andrew Hennessey, Eric Smith, Gary Swift, Nigel Kitching, Mark Ruffle, Q2A Media, Jonathan Edwards, Stephen Elford, Sim Marriott, Andrew Painter, Fred Blunt, Emma Brownjohn, Tom Percival, Dale Sullivan, Tom Cole
Cover design by Pearson Education Limited
Cover illustration by Volker Beisler © Pearson Education Limited
Printed in the UK by Scotprint

Acknowledgements
The Publishers would like to thank the following for their help and advice:
Liam Monaghan
Hilary Keane
Stephen Walls

Every effort has been made to contact copyright holders of material reproduced in this book.
Any omissions will be rectified in subsequent printings if notice is given to the publishers.